A STEP-BY-STEP BOOK ABOUT
FERRETS

JAY AND MARY FIELD

Photography:
Glenn S. Axelrod; Isabelle Francais (with cooperation of Parrots of the World, Rockville Centre, NY) Michael Gilroy, Mervin F. Roberts, Linda Smothers, Wendy Winsted.

Humorous drawings by Andrew Prendimano.

About the Authors

Jay and Mary Field live in Pittsburgh, Pennsylvania with their two ferrets, Furry and Heather. Five months after they bought Furry they formed the Ferret Fanciers Club to be a service to ferret owners, breeders and health care professionals.

Dedication

To all ferrets who bring their owners as much joy as do our two—Furry and Heather.

Distributed in the UNITED STATES by T.F.H. Publications, Inc., 211 West Sylvania Avenue, Neptune City, NJ 07753; in CANADA to the Pet Trade by H & L Pet Supplies Inc., 27 Kingston Crescent, Kitchener, Ontario N2B 2T6; Rolf C. Hagen Ltd., 3225 Sartelon Street, Montreal 382 Quebec; in CANADA to the Book Trade by Macmillan of Canada (A Division of Canada Publishing Corporation), 164 Commander Boulevard, Agincourt, Ontario M1S 3C7; in ENGLAND by T.F.H. Publications Limited, 4 Kier Park, Ascot, Berkshire SL5 7DS; in AUSTRALIA AND THE SOUTH PACIFIC by T.F.H. (Australia) Pty. Ltd., Box 149, Brookvale 2100 N.S.W., Australia; in NEW ZEALAND by Ross Haines & Son, Ltd., 18 Monmouth Street, Grey Lynn, Auckland 2, New Zealand; in SINGAPORE AND MALAYSIA by MPH Distributors (S) Pte., Ltd., 601 Sims Drive, #03/07/21, Singapore 1438; in the PHILIPPINES by Bio-Research, 5 Lippay Street, San Lorenzo Village, Makati Rizal; in SOUTH AFRICA by Multipet Pty. Ltd., 30 Turners Avenue, Durban 4001. Published by T.F.H. Publications, Inc. Manufactured in the United States of America by T.F.H. Publications, Inc.

Contents

We're glad you could join us. You are about to enter the new, wonderful world of the ferret fancy. Your owner-ship will give you some-thing in common with the people of ancient Egypt, where ferrets were chas-ing and catching mice (be-fore cats ever thought mice were a good meal!), as well as with your contemporaries, because ferrets helped the BBC to televise the wedding of Prince Charles and Lady Di to the whole world and also aided in "wiring" the United States' achievements in the space age.

INTRODUCTION

We strongly advocate obtaining knowledge about these animals, because a pet ferret is like any other living crea-ture. We hope you purchased this book before you acquired your first ferret, or shortly thereafter, if you've already taken one into your home. The ferret has certain requirements and you, as an owner, MUST accept complete responsibility for its health and well being. Your pet is, simply put, a focal point for you (and your family's) love and affection. A ferret is not a watchdog or a hunter, so you can "spoil" it to your heart's con-tent (with some exceptions). While your ferret certainly looks like a hunter with its alert, curious nature, long sinewy body and strong, short legs, it's not one. If released in the wild or accidentally lost, the animal will suffer an agonizing death due to starvation.

Since we've mentioned it, the other extreme of starva-tion is a proper diet for your ferret which will provide adequate nutrition for his or her growth and development. Failure to es-tablish the necessary nutritional requirements early in your fer-

FACING PAGE:
Ferrets are intelligent, playful, alert, and are
excellent pets!

ret's life may adversely affect its future health and development. Consider proper diet as your primary responsibility to your pet ferret. It's the same as if you were a new parent. You wouldn't feed a newborn baby tacos, would you?

Adequate housing is your second responsibility. Your ferret should be provided with a warm, comfortable and secure environment. If you've already purchased a ferret, we hope you're not keeping it in a shoe box or a small carrying case you may have bought at a pet store. If so, your pet may feel somewhat claustrophobic. Temporary housing can be a cardboard box at least 16″ × 18″ and over 20″ in height, covered with at

If you value your animal, you must provide it with secure quarters—ferrets can escape through seemingly impossible holes.

least two inches of cedar shavings or shredded newspaper. Other necessary additions include a heavy bowl for fresh water and another one containing a ferret feed or a good commercial kitten feed, and a small bath towel for the ferret to cuddle up in. The box should be kept in a comfortably warm area and be free of drafts. *Keep in mind that the shelter we described is strictly temporary.*

After establishing your ferret in its temporary quar-

ters, let it "catch its breath." After all, for a little animal the trip from the pet store or breeder, meeting new people and experiencing new surroundings, is quite an exhausting event.

If you have other household pets, keep the ferret separated from them for the moment.

After seeing your ferret comfortably installed in its temporary quarters, you can continue expanding your knowledge by reading and learning about the new, wonderful world of the ferret fancy. Who knows—at the end of the journey you might be raising a future Grand Champion or Grand Premier ferret!

We guarantee one thing—once you've owned your ferret you'll see why the name "ferret" came from the Latin "furritus," which means "little thief." These adorable little animals will indeed steal your heart.

Fed properly on a high-protein diet, ferrets will reward you with their condition, being sleek, fit, and shining with health.

Ferrets are ideal pets for a large segment of the population, with a few exceptions. Because they are clean, quiet, affectionate small animals they fit into lifestyles of many people better than dogs and cats and certainly offer more affection than tropical fish.

PURCHASING

They incorporate the best features of cats and dogs and eliminate some of the undesirable traits of these animals. They are like kittens—perpetually playful, but they don't claw the furniture or turn into felines who are extremely independent. They don't bark, knock over garbage cans, need to be walked in sub-freezing temperatures or growl at your guests.

There are just a few groups of people to whom we do not recommend adopting ferrets. If you travel extensively and cannot take your pet with you, you may want to wait until your circumstances change. Ferrets crave and thrive on your company and attention. Also, since these animals are so small and quick they might not be the best pets for elderly persons or those who are troubled by impaired agility.

Also, if you have children under six years of age you may want to delay this purchase until the children are a bit older. Ferrets, like other domestic animals, should not be left unsupervised with children. Small people can cause harm to these animals by accidentally stepping on them or neglecting to close a door or window.

Before purchasing a ferret, check your state and local restrictions. Unfortunately, there are some states and municipalities which ban ferrets (the FFC is working to eliminate these restrictions). Some towns require licensing of ferrets, so check this too. Currently California, Georgia, Massachusetts and New

FACING PAGE:
Selective breeding is producing new color varieties among ferrets. Sable and albino ferrets are shown.

Hampshire are states which still have laws which prohibit the ownership of ferrets. New York City requires an owner to pay a registration fee for ferrets.

PREPARING FOR YOUR FERRET

All clear on how your ferret will fit into your lifestyle, state or city? Well, don't run right out to purchase your little pet yet. You have an important duty to perform before it enters your home. That is FERRET-PROOFING it. Ferrets can get into incredibly small places. Our 2¼ pound male crawled into an

The classic example of the domesticated ferret in Queen Victoria's time was a white or yellowish pink-eyed albino.

uncovered space (2½" × 4") in our wall which was available for the installation of a master TV antenna! Arm yourself with plywood or thick cardboard, duct tape and a flashlight. Carefully check all surfaces under sink, stove, refrigerator, and continue this process in every room in which you will allow your ferret to have freedom.

House all secured? Now you're ready for the fun of choosing your little pet who will be a very large part of your

Ferrets respond well to affection and derive great pleasure from being stroked and petted when they are tame.

family for many years to come. The lifespan of a domestic ferret is from 8–12 years.

SELECTING YOUR FERRET

The obvious place to look for a pet ferret is at a pet shop, because even if a pet shop near you doesn't stock ferrets for sale, they can probably obtain a good one for you more quickly than you could yourself.

A ferret, in order to become a good pet, must be produced by a reputable breeder, since these are animals which require socialization to become pet stock. Breeding facilities must be USDA approved, and as a service to breeders and consumers, the Ferret Fanciers Club will approve breeding facilities and give a seal of approval to breeders who meet our requirements.

The minimum age for successful adoption of a kit is six weeks of age. Unlike kittens or puppies who leave their litter mates, they do not whimper or seem to miss their former family. If you do not plan to breed your ferret, you may wish to buy one that has been neutered or spayed and descented prior to sale.

Your pet will seem very small at the age of six or eight weeks, but don't expect it to ever achieve great size. The males will grow to approximately 21–24 inches long and the females 16–18 inches long from the tip of nose to tip of tail. Their adult

weight ranges from 1¼ pounds to 5 pounds. The males weigh more than females, and unneutered animals have higher weights.

If you are buying a neutered animal, there is little difference in temperament between the male (hob) and female (jill).

Take your time when selecting your ferret. First, check

Although ferrets do not have very good long-distance eyesight, their glittering eyes do not miss much and, of course, they do have an excellent sense of both hearing and smell.

the breeding facility or store. Is it clean and odor-free? Pick up the ferret you are interested in. Ferrets sleep deeply, so give the animal a little time to waken. When it is roused, it should be alert and curious. Its fur should be glossy and soft. Be sure there are no bare or scaly patches on the skin. Teeth should be clear, white, unbroken. Check to see none are missing. Heads

Regardless of color or fur type, all domesticated ferrets are scientifically designated as *Mustela putorius*.

of the males are rounder and broader with a shorter, tapered nose. The females' heads are smaller with a slightly longer tapering towards the nose.

The ears should be clean. The nose is doglike in shape and should feel moist if the ferret is awake. If awakened from a sound sleep, the nose should feel moist in a few minutes. Nose moisture is reduced during sleep because the ferret usually nuzzles its nose into its fur or bedding. The nose should not be cracked or scaly. Paws should be strong, without cracks or scaliness in the webbing or on the pads.

If the ferret has not been descented, there might be a slight musky odor about the animal. This can be corrected by the removal of the scent glands during the neutering or spaying procedure.

BRINGING HOME

In the large majority of cases, ferrets get along with most other household pets. To make this meeting easier, you might want to try this simple procedure.

The first night the ferret is in your home, let it sleep on a soft old towel. Let your other pet(s) sleep on old towels, also. The next morning, switch the towels so that the ferret is now sleeping on that of the other pet(s). This will enable the animals to become familiar with each other's scents, and they will feel like old acquaintances when they meet. If possible, let them meet on "neutral" ground—not the ferret's area or the sleeping or eating area of the other pet. SUPERVISE THEM WHEN THEY ARE TOGETHER. Let their time together be pleasurable—provide each pet with a treat. This will encourage the animals to associate pleasure with being together. Don't let them be together without your supervision until you're 100% convinced they'll get along. After all, a 20-pound dog is enormous to a small ferret.

After meeting household pets, let the ferret have some time to investigate any areas of your home you will open to it. We guarantee you'll have a great time watching these curious little things investigate. They will go around the walls of each room, sniffing every inch of the way.

If the ferret should start running around wildly or back up in a corner, it's time to return it to its litter pan—pronto!

If you are planning to allow your pet to run freely in all or most of the house, it is desirable to use more than one litter pan. The ferret will usually not make the effort to travel a great distance (to him) to find a litter pan. To encourage good toilet habits, keep the animal near its litter pan until it has performed after eating or sleeping. Accidents will happen, but you

FACING PAGE:
When handling your ferret, make sure you have a good firm grip (not too tight), as they are pretty "slippery."

Putting a bell on a ferret's collar can really be a life-saving device. Sometimes your ferret can't let you know it's in a compromising position.

can minimize them. If you catch your ferret in the act of making a mistake, pick it up and place it in the litter pan. When it performs, praise it lavishly or reward it with a treat. If it has already made the mistake, don't bother correcting it—it won't remember what it is being punished for. Simply clean the spot of the accident thoroughly to remove any odor, so the ferret won't return to the scent of the crime. Try using a mixture of ¼ white vinegar and ¾ club soda to remove any stains and odors. This combination is also very effective in removing other types of pet accidents and general stains.

If your ferret is a frequent offender (of not using the pan) try placing an additional litter pan in the place (usually a corner) the ferret has been using. No, it's not an ideal solution, but it will help in cleaning up life's little messes.

Keep your pet's food and water dishes in the place you choose; don't move them. The ferret will soon learn to go there for refreshments.

TAKING YOUR FERRET HOME

When getting out to purchase your pet, take along a small box such as a shoe box with holes punched in the top and a soft towel or blanket so the ferret can have a comfortable, safe trip to its new home.

Very kittenish in its play, a ferret will amuse itself for hours with a ball—jumping, bouncing, and pouncing.

You will want to get a permanent home ready for it to become accustomed to before much time has elapsed.

To facilitate litter training it is essential to confine the ferret to a small area (a large cage or a small, ferret-proofed room). If you opt for a large cage, be sure it is large enough to provide space for play, exercise, litter pan, feeding and sleeping quarters. A good size is approximately 14″ wide by 24″ long and

Ferrets love attention and food (not necessarily in that order).

10″ high. Wire cages are easy to clean. Be sure the ferret's exercise area is covered with plywood or other material so the animal won't be walking on the wires. The practice of walking on the wire is harmful to its paws.

The litter pan provided should be low in front with higher sides. Fill it with cat litter. Ferrets relieve themselves after a nap or eating and are unwilling to mess their sleeping or eating areas. This means they have no choice but to perform in the litter pan, which promotes good toilet habits.

In a cage, it is wise to use a heavy water bowl which can't be up-ended or a water bottle which attaches to the side of the cage. The food dish should be heavy and balanced well enough to ensure it is not knocked over.

Bringing Home

When purchasing your ferret, inquire what it has been eating. Buy some of this food so the ferret won't have the experience of getting used to new surroundings and new food at the same time.

If you decide to restrict your pet to one room to start, set up the supplies in the same manner (litter, eating, sleeping area). A good bed can be made from a plastic dishpan with soft blankets or towels. Bedding should be washed regularly. Do not use a cardboard box for a bed, as it can't be cleaned.

Once the ferret has become accustomed to the litter pan, do *not* completely change the litter material. Leave a small amount of feces in the new litter to remind your pet of the purpose of the pan. After the animal becomes more accustomed to the pan, you can use completely fresh litter with no problems.

Leave dry food out for the ferret at all times, and be sure there is constantly a supply of fresh water. Ferrets have very small stomachs and even a short wait for food can seem very long to the animal.

After your pet has rested a bit from the trip home and finished exploring new surroundings (usually one day) you can start to expand its world and let it truly become acquainted with the rest of the family.

Ferrets love to crawl into small, dark holes and can sometimes be found curled up asleep in the oddest places.

errets love to romp, run and jump like whirling dervishes. At full tilt they'll rapidly run backwards, tumble over, shake their heads back and forth vigorously and bare their teeth. This is all bluff and a lot of play.

PLAYING AND HANDLING

Our ferrets, Furry and Heather, have given us a merry chase around our home and then expect us to do the same as they scamper under, over and around the furniture with us in hot pursuit. All the while the ferrets are happily chittering to themselves or uttering a soft, guttural noise like "Chook, chook." It seems they're trying to see who wears out first—usually we do!

Another favorite game is "hide and seek." We go looking for them. One or the other will be hiding behind something, with their tails exposed. As we creep closer the tails will suddenly swish back and forth very fast. That's when we say "Gotcha" and they'll run to a new hiding place. We get as much of a kick out of it as they do.

Ferrets will usually play a short time with small-sized cat toys. These include soft squeaker toys, cloth covered mice, tiger tails with bells and a carpet-covered cat scratching post with a mouse and bell attached. Many ferrets enjoy some human baby toys (especially the kind that gives out a musical sound when pushed). After playing with their toys, ferrets will always try to hide their favorite ones; any mother will love the way they put their toys away.

To liven things up, we tie a string around the squeaker toys, the tiger tails and the cloth covered mice and drag the toy along the floor for the ferrets to chase. You can also dangle the

FACING PAGE:
Ferrets love toys that move, roll, bounce, and squeak, but some of the toys that fall into those categories are unsafe, as they break up and can be swallowed in whole or in part by a ferret.

Ferrets will usually get along with your other pets. Introduce them slowly and be ready to referee "just in case," but usually they will become fast friends.

toy just out of reach and watch the ferret jump for the toy. This string method allows you to animate the toy, which will create additional interest in the toy and will hold the ferret's interest for longer periods of time. Ferrets get so involved with their now-animated toy they will bite at it, fight with it and in some instances grip the toy so tightly in their mouths that you can literally lift them bodily off the floor (never more than a few inches). They just won't let go!

Anything can become a potential plaything for your ferret. Try the following: ping pong balls, cut-out portions of a plastic milk jug (1 gallon), large open-mouth plastic jars (minimum 3-inch opening), empty egg cartons and small cardboard boxes. Again, to liven things up, tie a string on them.

Another amusement that is easily obtained or which you might already have is a rounded or squared laundry basket (with an open lattice or cross-hatch). Ferrets enjoy exploring the insides of these articles, rolling around in them or just plain

chewing or scratching on the basket's plastic bars. You may want to try this: invert the basket over the ferret and watch it try to get out; they most always will if the basket is not too heavy. Now do it again. Watching closely, you'll see the ferret use the same procedure to escape from the basket as the last time. That's a learning experience. You'll find your ferret is a highly intelligent animal.

Just about anything is a potential toy for a ferret...as long as it's safe.

Another fun plaything is a flexible plastic exhaust pipe for clothes dryers. They are approximately 7 to 8 feet in length and 4 inches in diameter and can be purchased in most hardware or do-it-yourself stores. *Caution:* at the end of the pipe you'll find a sharp piece of wire. The wire should be covered with a strip of adhesive tape so it will not injure the ferret. Your ferret will delight in running in and out of the pipe, hiding in it,

and you can play too by slowly lifting the pipe, causing the ferret to slide out the other end. (Take it easy, though.)

Everybody has an old bath towel around. Just drag it along and your ferret will chase it, bite it and roll around in it, or you can roll the ferret up in it and just have some gentle roughhouse together.

Just about anything (provided it is safe) can become a plaything for your ferret—use your imagination and your ferret will do the rest.

Our two ferrets like it when we slide them across our bear skin rug or on a hardwood floor. This is done by taking the ferret, laying it down on its back and gently sliding it across the floor. They immediately jump back up and come back for more while chittering or chooking like mad. Use a highly polished floor so the ferret will slide easily, and make sure it is free of splinters.

HANDLING YOUR PET

From the start, handle and pet your ferret as much as possible. They respond to loving and cuddling, and this contact

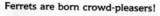

Ferrets are born crowd-pleasers!

When ferrets get excited or frightened, they arch their backs, fluff up, and make unmistakably angry noises.

with you helps them become better pets. Never grab for the animal. Approach it, speaking quietly. Put your hand on the floor and let the ferret approach you. Speak softly, then lift it, being sure to support the body along your arm. If it struggles to get down, hold on for a few seconds more, just to let it know that you're in control.

It is likely that your new ferret will bite. Don't take this personally—ferrets bite their litter mates, who, because of their thick skin, don't mind at all. However, our skin isn't thick and humans don't want to be bitten. This habit is easy to stop. Whenever your pet nips, thump him with your forefinger on his nose and say "no" in a very firm voice. Very little physical force is necessary—the key is to be consistent. You'll be surprised how quickly your pet catches on. When it is being held and *not* biting, be sure to reward good behavior with soft words, pets and small treats. The biting problem should end quickly.

In extremely rare instances a ferret will not overcome the biting problem. This is usually because the animal was not good pet stock. Once again, we emphasize, *buy from a reputable dealer.*

C uriosity is a trait we, as humans, admire greatly. Without it we wouldn't have had a Leonardo daVinci or gone to the moon. However, a ferret's instinctive curiosity can sometimes lead to its demise.

A ferret loves to explore every niche and

FERRET CURIOSITY

cranny in its new home. Normally, we'll smile and say, "Isn't that cute" when the ferret pokes its head out of a closet it has gotten into. Seriously, though, some members of the Ferret Fanciers Club have called, crying over the untimely death of their ferrets when the ferret's curiosity has gotten the better of it. One case was when a ferret had climbed into an open dishwasher. The owner turned it on, and the pet died. Another sad time was when a ferret slipped unnoticed from its home and was crushed to death under a bull's hooves. Still another almost tragic episode occurred when a pet ferret climbed, undetected, inside a refrigerator. All of a sudden, the family heard an unholy racket coming from their refrigerator. Everyone stared at this noisy appliance, thinking something must be mechanically wrong with the unit. The owner's son cautiously opened the door . . . there was the ferret happily exploring the interior, with predictably disastrous results. Egg cartons had been pried open and some of their contents were broken. The ferret by now was playfully batting the other unbroken eggs around the refrigerator's shelf as if they were ping pong balls.

The case of "How the Ferret Got Inside the Refrigerator" was solved when the owner's daughter admitted she had opened the refrigerator door 15 minutes earlier to get a glass

FACING PAGE:
It seems as if every ferret born wants to be somewhere other than where it is at any given moment.

Without fail, ferret-proof your house. Ferrets are natural hunters and feel compelled to stick their noses where they don't belong.

of orange juice and had not seen the ferret slip inside. Fortunately, the family had not gone out for an extended period of time or they would have come home to a dead ferret or one suffering from severe hypothermia.

Ferret owners have searched for their pets only to find them behind walls and panels, under or in refrigerators and dishwashers, inside closets and cabinets. One adventurous little guy had even climbed into an unconnected 3-inch PVC drainage pipe that a plumber had installed and planned to connect the next day!

To prevent such occurrences, we emphasize that you ferret-proof your house or apartment. All it takes is a stiff cardboard carton, cut up, folded and taped, if necessary, to block openings around and under the refrigerator, dishwasher, dryer and cabinets. If your ferret persists in scratching at the cardboard obstruction, spray some ferret repellent (ask your pet store) on it and say "NO!" while the ferret is scratching at it.

Ferret Curiosity

This should discourage further attempts. If not, repeat the process.

Also, if you have a mail slot, tape it shut and purchase a mailbox. The Ferret Fanciers Club has had many reports of ferrets who climbed out the mail slot and were lost or injured. Other precautionary steps you should take are:

1. Purchase a cat collar at your local pet store and attach to it a bell and a name tag. The bell will let you hear your ferret if it has gotten into one of its hidey holes and the name tag is essential if your ferret gets loose. Be sure the name tag includes your address and phone number.

While the literary-type ferret may be fascinated by your bookcase, it's not a good place to let ferrets play. Miraculously, these books are still on the shelves.

2. Buy a squeaker toy at your local pet store. It is an excellent calling device because ferrets are strongly attracted to its noise. To reinforce the squeaker's calling effectiveness, each time the ferret comes to you when you squeak the toy, reward it with a treat. Now your pet will come to you, rather than your having to search for it. This takes on additional importance if your pet has fallen asleep in some out-of-the-way place and you can't hear the attached bell. If it has become trapped in some confined area, you will hear its bell and its scratching to get out.

3. Watch out for those self-closing doors available on most refrigerators. They can slam shut with some force. If you have opened the door and your pet is checking out the goodies by standing on its hind legs and sniffing the bottom shelf, it might be hit by the closing door, possibly breaking some bones or causing severe internal injuries. Just exercise some caution if your ferret is up and around.

4. Never do the wash, take out the garbage or throw out empty bags without checking thoroughly, because ferrets *will* and *do* get into everything and anything.

5. Remove from your ferret's access all materials which, if ingested, could be fatal. These materials include cleaning products, insecticides, paint, solvents, medication, etc. Remember, if it can kill you, it will kill your ferret.

6. Check out your reclining chair before you stretch out in it since ferrets love to explore the chair's innards. Also, if you have a rocking chair, be sure your pet is not under the rockers.

7. Watch your step. Ferrets love attention and food. If your ferret places its front paws on your leg or foot, or if you are in the kitchen and your ferret is looking for a treat (both of ours usually are—that's why they are nicknamed "Foodies"), you're liable to step on your ferret. The screech they let out will scare the socks off you!

Having a ferret does place additional responsibilities on you and other members of your family. Don't leave open doors to your house or apartment, oven, dishwasher, washing machine, dryer, refrigerator or cabinets or you may have a lost, baked, washed, dried, well chilled, frozen or poisoned ferret.

Ferret Curiosity

FERRET LOST

In spite of the best of intentions, ferrets do get lost. If this should happen, don't panic. Take the squeaker toy that your pet will respond to and, while gently squeezing it, circle a small area where you last saw your pet. If this brings no results, increase the area while softly calling and squeaking. If the ferret does not respond, put its food and a favorite towel or blanket by your door. When the ferret has tired of exploring, it may return to rest.

Rabbits are the natural quarry of ferrets, as are rats and other small animals.

If your ferret should escape outside, a name tag on the collar with your phone number will greatly increase the chances of its safe return. Also, it is helpful to let your neighbors see your ferret (or least its picture). Let your mailman, delivery persons, etc., know you have a ferret. Many of our members' ferrets have been returned because neighbors knew they owned such a pet.

If your ferret is not recovered by searching the neighborhood, don't forget to check the local humane society shelter. *GO IN PERSON*—some of the staff may not know what a ferret is. Also, the Ferret Fanciers Club maintains a 24-hour hotline. If

These young, hand-raised ferrets are tame and docile and will make superb pets.

you've lost or found a ferret call (412) 322-1161. We've been able to return many ferrets to their happy owners through this hotline.

Our ferrets met their new neighbors in a unique, fun way. When we moved to our new home we were deluged with questions from neighborhood children—"What are they? Do

Ferret Curiosity

they bark? How big will they grow?" etc. To answer these questions and allow the children to meet the ferrets we held a birthday party for Furry, who was two years old shortly after we moved. We served the human guests ice cream and cake, and Furry and Heather had treats. Unfortunately, the guest of honor got bored with the festivities and fell asleep before the guests left. One of the enterprising little guests made up his own etiquette for the occasion—he reached down and shook Furry's tail and wished him happy birthday.

Ferrets are cuddlers. They are very content to curl up together and take a nap.

HEALTH

Shortly after you have brought your new pet home is the time for finding and visiting a veterinarian. Not all veterinarians treat ferrets, but it will be easy to locate one who does by calling to inquire before the first visit or by consulting the Ferret Fanciers Club, which will give you the names of the veterinarians in the area who we know treat these animals.

We recommend your becoming acquainted with your veterinarian before your pet is ill or there is an emergency. Your first responsibility as a pet owner is to insure your pet's health; to do this you need the assistance of a veterinarian.

The veterinarian will examine your ferret, weigh and measure it and give the immunizations that are necessary. When you purchase your pet, you will be told what (if any) shots it has received. Take this information with you. Bring a stool sample to be checked for parasitic worms. Also, if you have any questions about something that seems abnormal or unusual, this is the time to discuss it with your veterinarian.

Ferrets are susceptible to rabies and distemper, which also infect dogs and cats. The ferrets can be vaccinated against canine distemper after they have reached ten weeks of age with a primary injection of a vaccine containing a high titer of modified live distemper virus. Many authorities recommend they also be vaccinated against feline distemper. Discuss this with your veterinarian. Distemper vaccinations must be repeated yearly or every other year per your veterinarian.

During this visit a rabies vaccination may be given. The rabies vaccine which the veterinarian will administer is a killed only/murine origin. Rabies vaccination/immunization must be renewed annually.

FACING PAGE:
You can trim your ferret's claws with a dog's claw clippers. It's advisable to wear gloves during this procedure.

During this visit with your veterinarian, you should discuss vaccination against parainfluenza, hepatitis and leptospirosis.

Ask your veterinarian how frequently you should bring in your pet for routine examinations and adhere to that schedule. In addition, watch for symptoms of ill health. These include loss of appetite, listlessness, watery eyes, broken whiskers, lumps or abscesses felt when you gently rub your pet, diarrhea,

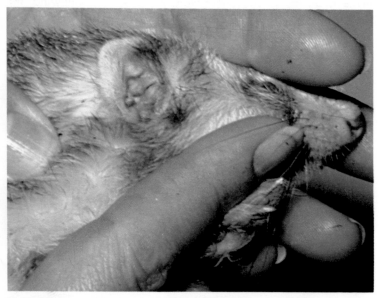

Your ferret's ears should be pink and free from wax and dirt.

blood in stools, discharge in the genital area, heavy wax in ears, prolonged scratching at ears, vomiting or sneezing and coughing which last more than two or three days. Also, if your ferret has been accidentally injured or has eaten a poisonous substance, DO NOT HESITATE TO SEEK MEDICAL CARE. The lives of many ferrets have been saved by prompt medical treatment.

After your sick ferret has been treated by your veterinarian there are a few things you can do to insure prompt, full

recovery. Of course, follow the advice of the veterinarian. Don't encourage the animal to engage in active play, but provide a warm, comfortable place for it to rest. Offer affection and cuddling, but if the ferret resists, allow it to rest comfortably on its own.

If medication is prescribed, be sure to give it to the ferret as long as it is advised by the veterinarian. Don't stop giving it simply because your pet shows signs of improvement. To make medicine giving and taking a bit easier, following are some tips:

1. Liquefy a pill and dispense it in a small oral syringe (available from your veterinarian or drug store) rather than hiding it in food or trying to force it down the pet's throat.
2. Try to medicate the ferret when it's already awake, rather than forcing it from a deep sleep.
3. Try to administer the medicine at the same time every day.
4. It's best to give oral medicine when the ferret has some food in its stomach. If given on an empty stomach, it may be thrown up.

A healthy ferret has bright, clear eyes, plenty of energy, and a good appetite.

5. If more than one kind of medicine is given, try to do all the medicating at the same time. If possible, mix it all together.
6. Wear a full apron or protective smock to prevent damage to clothing.
7. Hold the ferret in your lap on your left arm, slide the syringe into the side of the ferret's mouth and push the plunger in quickly. (If left handed, lay the animal on your right arm.)
8. Try not to allow the syringe to touch whiskers. This seems to alert the ferret that medicine is on the way and he will struggle.
9. Follow the medicine with a good-tasting treat and then gently play with or cuddle your pet. This will make it forget the unpleasant experience.

FERRET NOURISHMENT

A good diet for ferrets consists of specially prepared ferret food (available in pet stores) or a high-quality dry kitten food (minimum 35% protein). Do **not** feed cat food, as this does not have the high protein content which is necessary for a ferret throughout its entire life. Watch the ferret occasionally when it is eating. It may simply be hiding its food under furniture, in other corners, etc.

While ferrets enjoy an unusual treat, make sure you stick to high-protein food as the staple diet.

Health

Ferrets feel secure when they are fed in the same place and at the same time every day.

Ferrets enjoy snacks, treats and variety in their diet just as humans do. You will be amazed at the variety of food they will enjoy. Our ferrets like such varied items as watermelon, shrimp, tomatoes and raisins. Other ferrets we've heard of enjoy cucumbers, carrots, raw or cooked potatoes, spinach and celery. All food should be well washed and given in very small pieces. Again, watch to see that your ferret is not hiding this food. If the ferret shows signs of loose stools, cut back on treats.

Ferrets also enjoy foods which can cause problems, like stomach upsets or diarrhea. Foods which are milk-based, such as ice cream, cheese and other dairy products, may cause difficulties. Cake, cookies, etc., are definite "no-no's" although we, like most ferret owners, have been guilty of giving a crumb or two to ours on such gala events as a ferret's birthday or adoption day.

We are glad you will be joining the many thousands of owners of ferrets. Each day with a ferret is one of fun, affection and delight. With good care, you and your ferret should have many happy years together.

Unfortunately, the day will probably come when you and your pet will be separated by death. If your aged ferret has come to the point where illness is causing suffering and pain, the humane and loving thing is to ease your pet to its death. Euthanasia is the recommended and painless method.

GROOMING

Ferrets are by nature clean animals, but they do need some help from their owners. If your ferret is kept in the house, it should not be necessary to bathe it more than every two weeks. We have heard of many ferrets who dearly love bath time; unfortunately our two are not among them. The first few baths they had were quite a hassle for them. Now they reluctantly tolerate baths. To make baths easier follow these tips. Fill a sink (not a bathtub) with lukewarm water. Gently place the ferret in the sink and slowly soak the coat. Don't make abrupt motions, and talk soothingly to the ferret throughout the bath. Hold the ferret out of the water and drain the sink. Work a *non-tearing baby shampoo* into the coat, working up a rich lather. Don't forget the tail! Thoroughly rinse the ferret with lukewarm water running slowly from the tap. Be sure to remove all residue of the shampoo. Dry thoroughly with a large towel. After towel drying you might want to use a hand-held dryer *set on low heat and power settings*. Keep the ferret away from drafts after bathing.

Just remember—don't bathe the ferret more often than approximately every 14 days—more often than this will cause the coat to dry and encourage flaky skin. Never use any other shampoo except non-tearing baby shampoo; it will damage your ferret's coat.

If your pet has been repeatedly scratching and biting various portions of its body, fleas might be the problem. To check for flea infestation, place your pet over a piece of white paper and ruffle its fur. The dislodged fleas will stand out against the background. Examination of the ferret's coat should show black deposits in the fur if fleas are present.

FACING PAGE:
Light brushings will help stimulate natural oils in
your pet's coat, keeping it clean and shiny.

Fleas are more easily seen on light-colored animals, resulting in early detection and extermination.

If fleas are evident, use dog/cat flea shampoo (avoid contact with the eyes). Thoroughly clean the ferret's bed and bedding materials. A cat flea and tick collar may also be used. *Be sure to allow the collar to air out 24 hours before putting it on your ferret.* Remove the collar during bathing.

There have been some reported cases of ferrets losing their fur due to flea collars. Should this occur, remove the collar immediately and rely on anti-flea shampoo.

To prevent torn nails and infection, the ferret's nails should be clipped once a month. Use a pet nail clipper. *Extreme care should be exercised when clipping not to cut the red colored section extending from the root of the nail.* To see it, hold the nail up to a bright light. Nail cutting is easier if done when the ferret is sleeping—this will prevent some struggling. If you are unsure about clipping your pet's nails, your vet will show you the proper procedure.

Grooming

Clean the ferret's ears monthly with a cotton-tipped swab which has been moistened with mineral oil. This will prevent the infestation of ear mites. Be very cautious when cleaning the ears—go around the outer ear area to remove wax. If there is severe wax build-up, take your pet to the vet.

After a grooming session, be sure to be lavish with praise, affection and rewards for your pet.

COAT CHANGES

Ferrets will shed their coats and grow new coats two times a year. These changes are meant to occur in the fall and spring, but ferrets who live in light-filled houses will not adhere strictly to that schedule. You can speed the shedding process by lightly brushing your pet. After the old coat has been shed a new one will appear (lightweight and shorter in spring and full and luxurious in the fall).

Even the indoor ferret will have a spring and fall wardrobe—lightweight for spring and full and heavy for fall.

THE FERRET ON THE GO

Ferrets can accompany you much of the time and will want to accompany you *all* of the time. They can be trained to walk on a leash with little trouble. Be sure the leash is a harness-type that can be adjusted to fit snugly, but not tightly. Practice walking your pet on the leash in the house. BE SURE YOUR FERRET CAN'T SLIP OUT OF THE HARNESS! Your pet will also love riding on the collar of your coat, on your shoulder or in a canvas bag with air holes. One of our members has a "corporate ferret." This couple own their own business and can take their pet to work with them each day. The ferret rides in a canvas shoulder bag (with mesh in one part of the front for air) to and from the office. While at the office he sleeps in his cage or walks around on a leash. His owners are not sure if customers are coming to see them or their ferret!

Ferrets travel well in cars—just be sure the windows are not opened wide enough for them to crawl out of. Don't leave your pet in the car if the temperature is over 70°. The heat builds up quickly and can cause illness or death. By the same token, be sure it's not too cold for your pet. When away from home, take a small amount of food and water in a feeder bottle. Also, a small pan filled with litter is handy.

Don't try to handle your ferret until it is completely awake. Even tamed, a sleepy ferret can be a cranky ferret.·

Grooming

Your animal will be a real attention-getter wherever it goes. However, it's not advisable for strangers to handle it for two reasons: your pet might be frightened and try to bite, and the strangers might have cold viruses which are readily transferrable to ferrets. "Look but don't touch" is a good rule for outings.

Your pet can travel with you on many trips. It is essential to have a carrying case whether you travel in a car, bus, train or plane. If you are using any other means of transportation besides your car, check with the transportation company about regulations.

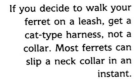

If you decide to walk your ferret on a leash, get a cat-type harness, not a collar. Most ferrets can slip a neck collar in an instant.

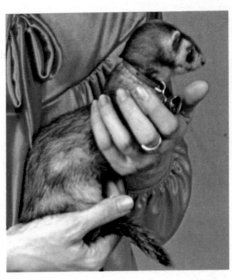

As we mentioned, there are restrictions against ferrets in some areas. To prevent problems, take along ownership papers and a statement from your veterinarian listing the inoculations your pet has received. If ownership papers are not available, have your veterinarian sign a note stating where you reside and the fact that you and your pet live at the listed address. (If you move to another state, take this certification along too.)

Many hotels and motels accept pets. Check this out before making travel plans.

To breed or not to breed? That question must be well considered. Breeding of ferrets is not as simple as breeding cats or dogs. Unlike cats and dogs, not only does the female ferret need to be in heat (estrus), the male must also be in season. Male ferrets who are not to be used for breeding should be neutered and descented.

BREEDING

This is a surgical process performed by a veterinarian, which removes the reproductive organs and scent glands located at the base of the tail. This surgical procedure is simple and requires only a short stay at the veterinarian's office. Often the ferret can return home several hours after surgery. Many ferrets are neutered and descented before being sold (prior to six or eight weeks), but early neutering will usually cause fading in the coloration of the ferret's face mask.

If the female is not to be bred, she *MUST* be spayed and should also be descented. A female is usually sexually mature at six months and will go into heat during the breeding season (usually from March through August). She will show her readiness to breed by a swollen vulva. If she does not conceive at this time, the increased estrogen level in her body can depress the red blood cell production in the bone marrow, causing aplastic anemia. This is always fatal if not treated in the early stages. The chances of uterine infection are also greatly increased during heat.

If your female goes into heat before she has been spayed, your veterinarian will give her shots to bring her out of heat. Once this has happened (in about 7–10 days) she can then be spayed and descented.

Unneutered ferrets are more aggressive than those which have been neutered. Unless you are absolutely certain

FACING PAGE:
A male ferret is called a hob, or dog. The female
ferret is called a jill. The young are kits.

The external gentalia of the female ferret.

that you wish to go into ferret breeding, neutering and descenting is the best way to go.

If after careful consideration you decide to breed, you will need to be informed about the process of breeding. Be sure your pet is of good pet stock.

A male ferret will indicate his readiness for breeding when his testicles become hard and large. If he is not in season for breeding, he will not perform with the female. If your male is ready to breed you can find him a suitable mate by advertising in your local newspaper, checking with your veterinarian for a female in heat or by putting a notice in the Ferret Fanciers Club newsletter.

Signs of heat in a female are a swollen vulva with a slight discharge. To find a suitable male, follow the same advice as given in the above paragraph. The female should be taken to the male. The act of mating is very rough—we know of one breeder who will not allow the owners of the female to watch the breeding as it seems hard on the female. To arouse her the male will bite her on the back of her neck. This may cause redness, hair loss or bleeding, but is natural for ferrets. The breed-

ing session may last as long as 45 minutes to 1½ hours. If breeding has occurred, the female's vulva will reduce in size within one or two days. At that time the ferrets should be separated.

A ferret's gestation period is about six weeks. A jill will usually want more to eat and want to sleep longer than usual. That is normal. Two weeks before the kits arrive, the female should be put in a warm bed with fresh nesting material. Let her eat as much as she wants. Check with your veterinarian re-

Jills are very good mothers and have incredible patience with their lively youngsters.

garding supplemental vitamins. Let the ferret become accustomed to her new bed, so she will feel comfortable there when her kits are born. Prior to birth it is normal for a jill to lose a considerable amount of hair.

When your ferret gives birth you should be there, but allow your animal to handle things. Be sure the female has food, water and, if necessary, help from your veterinarian, but *do not handle the kits*. Above all, do not remove the afterbirth. This must be eaten by the jill. If she does not perform this function, the vital hormones in the afterbirth and placenta will not trigger the lactation stimulus and the mother will not be able to nurse. If your ferret shows any signs of distress or inability to deliver the kits, call your veterinarian immediately.

On rare occasions a mother giving birth will completely ignore her kits or she will give birth to a very large number of kits (sometimes as many as 12–15). Since the mother has only eight nipples, some kits, due to limited milk supply, will become weak and die. To prevent unnecessary kit fatalities, you should try to locate, prior to birthing, another female who has recently had kits and may still be nursing. This female could be the wet nurse for the rejected kits.

Sound asleep until the next meal! At this age, like human babies—it's eat and sleep, eat and sleep.

Introduce kits to their surrogate mother using the "Mary Fetter Method." While wearing white cotton gloves, thoroughly handle kits and introduce the gloves to the surrogate mother ferret.

Check the nest every day to be sure that none of the kits have wandered or been pushed away from the mother and have died. Be cautious when approaching the nest, as the mother is very protective of her kits and may bite you. Do not allow the male to be with the mother or kits.

Breeding

From the time the kits are 27 days of age they must be socialized by humans. Their eyes will still be closed, but they are not too young to know you are a human and that it feels good to be cuddled and petted. This socialization is the key to raising a loving, affectionate ferret which will be a good pet. *Do not neglect this vital aspect of their rearing!*

At approximately six weeks the kits can be weaned. To do this, introduce a bowl of wet feed (regular ferret food soaked with water).

Your responsibility to the kits includes finding good homes for the ones you do not keep. Contact your veterinarian, who may know of individuals wanting a ferret, put an ad in the paper, or place a notice in the Ferret Fanciers Club newsletter. Be sure prospective owners are aware of proper treatment of a ferret, and be available to answer questions. It might be a good idea to provide each purchaser of a ferret with a copy of this book, the cost of which can be included in the purchase price.

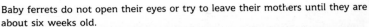

Baby ferrets do not open their eyes or try to leave their mothers until they are about six weeks old.

Ferret shows are a great way of meeting other ferret owners, checking to see how your particular animal meets general and specific standards for ferrets and, most importantly, you will have a chance to show off the world's most superior ferret(s)—YOURS.

SHOWING

The Ferret Fanciers Club sponsors shows where ferrets can compete in a variety of classes. The classes are:

Coloration—Breeding Ferret
Conformation—Breeding Ferret
Coloration—Non-breeding Ferret
Conformation—Non-breeding Ferret
Junior—for a kit 14 weeks to one year of age

At a ferret show you may also enter your pet in a "fun" class such as "Best Dressed." In recent Ferret Fanciers Club shows held in Pittsburgh, Pennsylvania, many of the ferrets in Best Dressed far outshone the humans, sporting such costumes as an angel or a Spanish bride.

To help you make your decision on entering a ferret show we have listed General and Specific Standards, Coloration Standards and the proper way to handle your ferret for judging.

There's lots of suspense as the final points are totaled up and the winners of each category are again called up to be rejudged before naming "Best of Best." Whether your pet wins an honorable mention or a ribbon or not, we guarantee you will have an enjoyable afternoon.

GENERAL CONFORMATION STANDARDS
(ACCEPTED BY FERRET FANCIERS CLUB)

Alertness—ferret or kit should exhibit inquisitive behavior. Lethargic behavior or listlessness should be penalized.

FACING PAGE:
Ferret clubs and shows help you to learn more about your pet. The more you know about ferrets, the better able you are to take care of them.

Condition—right amount of flesh, not excessively fat or thin. Coat is full and healthy and shows good care.

Balance—well proportioned: i.e., head not in proportion to body, oversized features or glaring faults should be penalized.

Soundness—free of any disabilities: i.e., lameness, missing limbs or lumps under the coat. Ferrets or kits showing these disabilities should be disqualified.

Note #1: Any ferret or kit who bites or exhibits aggressive behavior must be disqualified.

Note #2: Any ferret or kit which fails to gain any points in the judging of General Conformation Standards must be disqualified.

SPECIFIC STANDARDS

Head—Males should have broader and rounder skulls, firm muzzle line with shorter and more abbreviated tapering to the nose. Females should have smaller heads, firm muzzle line, with a slightly longer tapering towards the nose.

Bite—Bottom teeth should fit firmly behind the upper canines. Severe undershot (lower teeth just forward of the upper canines) or poorly aligned jaw line (wry mouth) should be penalized cuts. Teeth should be clean and white (no discoloration), and unbroken. None should be missing.

Profile—Ferrets and kits should have, when standing on all fours, heads poised in an alert manner with a slight arch in the middle of the back. Guard hairs should be erect.

Ears—Should be short, well-rounded and close to the skull. Ear opening at the base should be broader. Coloration at the tips of the ears should match the undercoat. Coloration at the base of the ears should match point color (tails, legs and mask). Sharp deviations are subject to cuts. Slight nicks in ears of female breeding ferrets may be ignored, but not cases involving severe mutilation. Ears should be free of debris.

Ear Placement—Ears should be evenly spaced, neither too close nor too far from one another.

Eye Shape—Small and round, well positioned between ears and nose. Overly small eyes should be penalized. Note— eye secretions may indicate infection.

Eye Color—Should reflect light and have color harmony with coat; i.e., black with darker coats, deep brown with lighter coats and pink to red in whites, EXCEPT onyx (black-eyed whites).

Breadth Between Eyes—Moderate spacing; close sets, wide sets or one eye higher or lower than the other deserves cuts.

Whiskers—Should be long and full. Short or broken whiskers may indicate poor diet or other medical problems and should receive cuts. Lack of whiskers or extremely short whiskers should be reported to the Show Committee.

Nose—Should be canine in shape and slightly moist. Surface color should conform to coat color: i.e., darker coats—darker noses. The lighter the coat, the lighter the nose. Cracks in or scaliness on the nose deserve cuts. Mucous discharges from the nostrils must be reported to the Show Committee.

Neck—Should be strong and muscular without folds or

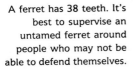
A ferret has 38 teeth. It's best to supervise an untamed ferret around people who may not be able to defend themselves.

laps. Neck extension capability should be approximately 1½ times its length when the ferret or kit is at rest. There should not be any narrowing of the neck at the head or shoulders.

Shoulders—Should show good definition and not exceed the base of the neck.

Body—Should be slightly elongated with smooth lines when still. In motion, body should be rounded without showing any sharp or flat angles.

Body Size—In breeding competition, ferrets should be

slightly larger than non-breeding. Males (3–5 pounds, 16–20 inches in length) are usually larger than females (1–3 pounds, 14–15 inches in length), but a large breeding female should not be penalized. Grossly overweight or underweight ferrets or juniors should be penalized.

Mask—Should surround both eyes and conform to point (legs and tail) color for ferrets in breeding competition. Junior neutered or spayed ferrets (non-breeding) usually lack a distinctive mask.

Tail—Should be straight (no curvature) with a thick base tapering to the tip and fully furred.

Legs—Should be short, heavy in bone and strong.

Feet—Paws should be well-furred between the pads. The paw shape should be oval and show no splaying between the toes. Pads and webbing should be free of cracks and scales.

Pads—Color may differ, from black, pink, spotted or mottled. All pads should be uniform in color.

Coat—should be free of scaliness or bare patches. It should be clean, soft and smooth and not contain any matted areas or appear to be excessively oily.

FERRET FANCIERS CLUB COLORATION STANDARDS/ DEFINITIONS

Coloration standards for adults (1 year and older) breeding, neutered and juniors are light sable, sable, hooded sable, sterling silver, silver mitt, ruby white, onyx white, butterscotch, cinnamon and experimental.

Sable—should be bold and uniform in color. A clear distinction between the guard hairs and underfur is especially desirable in sables. Guard hairs should be dark brown but of a close hue. Underfur can range from almost white to a golden yellow, but with evenness being all important. The mask on sables is the most important color point. It should be pronounced and evenly balanced and positioned on the face. A vivid raccoon type mask scores highest. Even and uniform markings on the extremities are next in scoring. This color is sometimes shown with four white feet and a bib of white.

Hooded Sable—the standards for sable apply to ranch sable except for the mask. The mask should appear as a hood

and cover the eyes with no light color on the forehead.

   Onyx-eyed White—should show an even body and head with no tipping or mask. Its special quality is the black eyes. Acceptable base color ranges from a pure white to a yellow, with white being scored higher.

The ferret is very agile. It twists and turns like a contortionist until one wonders if it will ever be able to straighten out again.

Ruby-eyed White—standards are the same as onyx-eyed except for eyes of a dark red. Eye color ranges from rose to ruby with ruby preferred.

Silver (silver mitt)—The silver should have a coat of guard hairs which is as close to black as possible with white hairs interspersed to give a silver appearance from a distance. This color phase should have a white bib and four white feet plus a white tip on tail. The mask on the silver is not as pronounced but should be even and symmetrical. The darkness of the mask is a plus in scoring, as is even length and intensity of feet markings.

Sterling Silver—a smooth and even sprinkling of black guard hairs on a white underfur give this ferret its name. Grading this color is the perception of an almost gunmetal appearance from a distance. Eyes must be black in this color variety. Mask carries almost no importance if it blends with the overall.

Butterscotch—this color phase is distinguished by an auburn or butterscotch appearance in contrast to the sable. Other standards are the same as the sable. In some parts of the United States this color is referred to as siamese. A subvariety of this color with four white feet and a white bib is known in many parts of the country.

Cinnamon—should appear as a white ferret with an even sprinkling of pale red guard hairs on the body with points being darker and solid. The mask should match the points.

Experimental—spotted or other color without the ability to breed true.

General Color—should conform to Ferret Fanciers Club recognized colors. Ferrets or juniors should be entered under proper color for show purposes. Guard hairs should conform to points (legs, tail and mask). Undercoat should conform to Ferret Fanciers Club recognized colors. Special note: silver-mitted ferrets or juniors must show white on all four paws or be penalized.

EXHIBITOR HANDLING FOR JUDGING

First Position: Exhibitor shall present the ferret in profile (legs down) with left hand holding the head and legs. The

right hand should support the hind legs with the tail positioned along the right forearm.

Second Position: Exhibitor will reverse the profile position; right hand holding the head and front legs, left hand the hind legs; tail positioned along the left forearm.

Third Position: Exhibitor will turn the ferret to a head-on position facing the judge. The judge may then ask the exhibitor to:

a. Open the ferret's mouth (place your thumb and forefinger on each side of the ferret's mouth and gently pry open the jaws).

b. Turn the head left, right, up or down using your right hand.

c. Place the ferret on the judging stand.

Win, lose, or draw, a ferret show is fun for everyone.

d. Gently blow into the ferret's fur covering the back from tail to head.

Fourth Position: Exhibitor will hold the ferret in a vertical position; right hand holding the head and legs; left holding the hind quarters with the tail hanging down with the ferret's abdomen (underside) facing the judge. This position will allow the judge to examine the ferret's:

a. abdomen

b. legs, toes, pads and claws

c. ask you to gently blow into the fur covering the underside from tail to head

d. ask you to gently spread apart the hind legs to expose the genitalia.

Exhibitor may be requested to hand the ferret to the judge for a more detailed examination.

Clubs

When we bought our first ferret in October, 1984, we knew of no other ferret owners. The staff of the pet shop was helpful with advice on food, training and general care. They were able to give us the name of only one veterinarian in the Pittsburgh area who treated ferrets. As ferret owners, we felt very alone. Two months later, at a holiday party, we met a gentleman who also owned a ferret. We're afraid we were not the ideal guests at that event—we did not "mingle." For the entire evening the three of us huddled in a corner exchanging ferret tales (sorry, couldn't resist that pun).

At that time we realized there must be many ferret owners in the country who would like to receive and exchange information or just communicate with other ferret fanciers. To help in this process we started the Ferret Fanciers Club in April, 1985. We were fortunate that a reporter from the *Pittsburgh Post Gazette* was fascinated by ferrets and he ran a large article. Shortly afterwards the Associated Press featured us in articles which ran across the nation. Since that time we have been interviewed by numerous publications, including the *Wall Street Journal,* which ran the article on ferrets on their front page in April, 1986.

Presently the Ferret Fanciers Club has members in 49 states (we are still waiting for a member from Hawaii) and Canada. We serve our members in the following ways:
* provide an exchange of medical information for health-care professionals on a monthly basis.
* publish a monthly newsletter.
* maintain a 24-hour hotline for lost or found ferrets, emergency questions or veterinary referral.
* offer charters to ferret clubs nationally and in Canada.
* set coloration and conformation standards for ferrets.

The Ferret Fanciers Club's purposes are:
1. To sponsor and promote the welfare of all ferrets.
2. To cultivate friendship among ferret owners, breeders and fanciers.
3. To hold and promote ferret shows.
4. To promote public awareness of ferrets as household pets.
5. To disseminate knowledge and information on ferrets.

Ferrets are tremendous escape artists and can exit through incredibly small openings. The ferret's quarters must be secure enough to contain these wonderful little beasts.

6. To insure high standards in the practices of breeders.
7. To assist owners and prospective owners as needed.
8. To work to legalize ferrets in all states.

To join the Ferret Fanciers Club, write FFC, 713 Chautauqua Court, Pittsburgh, PA 15214 or call (412) 322-1161 to obtain information about membership, registration and application form.

The following books by T.F.H. Publications are available at pet shops everywhere.

FERRETS AND FERRETING—
By Graham Wellstead
ISBN 0-87666-938-0
PS-792

The author draws on his years of experience to provide insights into ferrets, with extensive coverage of all aspects of keeping ferrets, breeding and hand-rearing. The information presented here will be interesting to readers who wish to know everything possible about this most curious of domesticated animals.

Hard cover, 5½ x 8", 192 pages

SUGGESTED READING

ALL ABOUT FERRETS—By Mervin F. Roberts
ISBN 0-87666-914-3; PS-754

All About Ferrets is a basic book that highlights vital information about this very popular animal. Chapters include topics on handling, training and taming, the proper way to choose a ferret, cleanliness, diet, health care, and reproduction. This book is completely illustrated with 21 full-color photographs. The information presented here is everything a reader needs to know to keep ferrets as pets.

Soft cover, 5½ x 8", 64 pages

FERRETS—By Wendy Winstead
ISBN 0-87666-930-5; KW-074

The information given in *Ferrets* is presented in an interesting and easy to read style. This book is illustrated with 33 full-color photos and 27 black and white photos. The chapters present sensible, easy-to-follow recommendations about selecting and caring for pet ferrets. It concentrates on providing readers with the information they need and want.

Hard cover, 5½ x 8", 96 pages

Index